SpringerBriefs in Energy

SpringerBriefs in Energy presents concise summaries of cutting-edge research and practical applications in all aspects of Energy. Featuring compact volumes of 50 to 125 pages, the series covers a range of content from professional to academic. Typical topics might include:

- A snapshot of a hot or emerging topic
- A contextual literature review
- A timely report of state-of-the art analytical techniques
- An in-depth case study
- A presentation of core concepts that students must understand in order to make independent contributions.

Briefs allow authors to present their ideas and readers to absorb them with minimal time investment.

Briefs will be published as part of Springer's eBook collection, with millions of users worldwide. In addition, Briefs will be available for individual print and electronic purchase. Briefs are characterized by fast, global electronic dissemination, standard publishing contracts, easy-to-use manuscript preparation and formatting guidelines, and expedited production schedules. We aim for publication 8–12 weeks after acceptance.

Both solicited and unsolicited manuscripts are considered for publication in this series. Briefs can also arise from the scale up of a planned chapter. Instead of simply contributing to an edited volume, the author gets an authored book with the space necessary to provide more data, fundamentals and background on the subject, methodology, future outlook, etc.

SpringerBriefs in Energy contains a distinct subseries focusing on Energy Analysis and edited by Charles Hall, State University of New York. Books for this subseries will emphasize quantitative accounting of energy use and availability, including the potential and limitations of new technologies in terms of energy returned on energy invested.

More information about this series at http://www.springer.com/series/8903

Flávia de Andrade · Miguel Castilla ·
Benedito Donizeti Bonatto

Basic Tutorial on Simulation of Microgrids Control Using MATLAB® & Simulink® Software

 Springer

Flávia de Andrade 🔘
School of Engineering and Physical Sciences
Heriot-Watt University
Edinburgh, UK

Miguel Castilla 🔘
Department of Electronic Engineering
Universitat Politècnica de Catalunya
Vilanova i la Geltrú, Barcelona, Spain

Benedito Donizeti Bonatto 🔘
Institute of Electrical Systems and Energy
Federal University of Itajubá
Itajubá, Brazil

ISSN 2191-5520   ISSN 2191-5539   (electronic)
SpringerBriefs in Energy
ISBN 978-3-030-43012-2   ISBN 978-3-030-43013-9   (eBook)
https://doi.org/10.1007/978-3-030-43013-9

This Springer imprint is published by the registered company Springer Nature Switzerland AG
The registered company address is: Gewerbestrasse 11, 6330 Cham, Switzerland

Preface

A microgrid is a system with distributed generation sources, energy storage devices and feeding loads, all integrated and controlled as a singular unit with defined electrical boundaries regarding the main grid. Microgrids can operate connected or disconnected to the power utility grid. Fundamental properties of enabling microgrids are flexibility and efficiency, which are associated with the power flow control toward the main grid. These properties are designed to provide a system that can respond to changes in load demand and distributed generation output in a timely and effective manner, without compromising performance and stability. In this regard, the design of microgrid control strategies must be robust to properly work along with utility power network and independent of it, sustaining the power quality standards.

Through case studies, this tutorial aims to facilitate the learning process of modelling and simulating control methods of power electronic converters, which are at the interface of distributed energy resources operating equivalently to current sources or to voltage sources. The development of this *Basic Tutorial on Simulation of Microgrids Control Using MATLAB® & Simulink® Software* was originally proposed by Benedito Donizeti Bonatto. Under the leadership of Miguel Castilla, Flávia de Andrade wrote the contents, modelled and simulated the case studies based on the knowledge gathered from various researchers of the Power Electronics and Control Systems (PECS) research group—https://pecs.upc.edu/—at the Universitat Politècnica de Catalunya (UPC).

This tutorial is organized as follows: first an overview of microgrid operating modes and control methods is presented and then five case studies are designed and simulated or proposed, on the MATLAB-Simulink software platform with various control techniques applied to microgrids on different operating modes. Hopefully, this document will assist the hard task of integrating theoretical and practical multidisciplinary topics in this field, as a mean of introduction to the subject and not full instruction. Whenever possible, links to videos on YouTube will also be available with step-by-step guidelines regarding the case studies simulations.

The tutorial was greatly supported by the Erasmus Mundus—Smart Systems Integration (SSI+) programme, Coordenação de Aperfeiçoamento de Pessoal de Nível Superior CAPES—Brazil—Finance Code 001, Conselho Nacional de Desenvolvimento Científico e Tecnológico (CNPq)—Brazil, INERGE and FAPEMIG. The authors also wish to acknowledge the Federal University of Itajuba (UNIFEI) and UPC, especially the PECS group members José Luis Garcia de Vicuña Muñoz de la Nava, Jaume Miret Tomas, Antonio Camacho Santiago, Manel Velasco Garcia, Pau Martí Colom and Ramon Guzman Sola, whose availability is greatly appreciated and work was used as reference for many steps during writing the tutorial and simulating all case studies.

Edinburgh, UK   Flávia de Andrade
Vilanova i la Geltrú, Spain   Miguel Castilla
Itajubá, Brazil   Benedito Donizeti Bonatto

Contents

List of Figures

List of Tables

Chapter 1
Microgrids: Operation and Control Methods

Abstract A microgrid is a distributed system configuration with generation, distribution, control, storage and consumption connected locally, which can operate isolated or connected to other microgrids or the main grid. It contrasts with traditional centralized grids through bidirectional connection with users and autonomous local control layers. Advances in control techniques, automation, energy storage technologies, communication and superior computing processing capabilities lead to increased efficiency and reliability of microgrids compared to traditional grids. Electrical grid control methods are divided into three hierarchical levels, diverging on the scale and purpose of its implementation. The most promising method so far is based on a technique known as "droop control", which was developed based on conventional power systems to support the parallel connection of multiple voltage sources sharing the network loads. This chapter introduces an overview of the electrical energy industry evolution, after decades of centralization, reoriented towards increased distributed generation and microgrids. Also, a classification of microgrid operation modes is presented, including grid-connected, islanded and transient operation mode. Finally, the chapter presents a comprehensive description of microgrid control strategies based on the classical hierarchical control method.

Keywords Microgrid · Droop control · Power inverter · Distributed generation · Control methods · Operating modes

1.1 Introduction

The electricity generation, transmission and distribution industry went through many different stages of development, with new technologies still evolving today. Its whole infrastructure is broadly referred to as electrical power systems. In the beginning of the 20th century, society moved towards the consolidation of previously fragmented power generation competitors, usually located close to their end consumers, into large scale state-sanctioned monopolies, a strategic move that was necessary to ensure the creation of standardized infrastructure and increase service reliability for a technology that in less than a century of its development became a basic human need. The implementation of central controls acting as dispatchers for these interconnected generation units triggered the growth in the field of electrical power systems engineering, which involves the deployment and optimization of computing facilities to create electrical power systems that are efficient, safe and reliable [1].

Due to the rising load demand and the pressure for environmentally friendly power generation technologies, the balance between distributed and central generation of electricity is being re-evaluated, with modern models of sustainable power and smart cities relying on grids with an increased number of low-capacity distributed generation systems installed [2]. In this scenario, distributed generation is referred to as small scale generation systems (typical capacity less than 50 MW) installed close to or even as part of the end consumer facilities, which can act independently of a central load dispatcher and are usually linked to renewable energy sources, such as wind turbines or solar photovoltaic panels powered systems [3, 4].

The rise in use of distributed generation systems enabled the formation of a local complete generation-distribution-consumption electrical power system, essentially composed of distributed generation units, energy storage devices and loads, effectively managed as a single controllable unit with distinct electrical boundaries related to the main grid. This configuration is referred to as a microgrid, and its truthful implementation carries not only advantages in environmental impact, but also increases local reliability and reduces risk through using a less capital intensive infrastructure [3]. A microgrid is characterized as flexible and intelligent, which can be arranged in different configurations, according to the level of integration with the main grid. These variations are called operating modes and can be either temporary conditions due to the current state of the grids or a design selection [5]. A classification of operating modes is presented in Sect. 1.2.

In order to achieve optimal grid performance and integration between the traditional grid with microgrids systems, the implementation of control techniques is required [6]. Control methods of microgrids are commonly based on hierarchical control composed by three layers: primary, secondary and tertiary control. Section 1.3 describes microgrid control techniques based on the hierarchical control method.

1.2   Operation Modes of a Microgrid

This section describes the main operating modes: grid-connected mode when there is an interaction with the utility grid; islanded mode referring to an autonomous operation; and transient operating mode, as stated by the name, it is the transition means when there is a disconnection or restoration in respect to the main grid [5].

1.2.1   Grid-Connected Mode

In the grid-connected mode, the microgrid operates by importing and exporting energy from and to the power utility grid, ensuring energy and power control flow balance and supporting the grid through an array of ancillary services, such as voltage and frequency control regulation, dispatch services and operational reserve capability, among others functionalities according to the load and generation conditions [3, 5, 7]. Ideally, microgrids are consistently interconnected to the utility, enabling any excess of energy from the microgrid to be sent to the main grid, as well as any deficit of energy in the microgrid to be supplied by the utility, which should be sporadically since the microgrid must be self-sufficient designed [8].

In this operation mode, the power utility grid regulates the network voltage amplitude, frequency and phase at the point of common coupling (PCC), corresponding to an electrical connection point between the microgrid and the main grid [2]. The converter systems, which are electronically coupled to distributed energy sources dedicated to one load or a group of loads, normally act as voltage followers in the grid-connected mode and are classified as *networking-feeding*. When power electronic converters are controlled as network-feeding, they are permanently synchronized with the main grid and are configured to operate equivalently as current sources. This operation is mainly implemented on non-dispatchable intermittent renewable energy sources that require maximum power point tracking (MPPT) algorithms, which is credited for extracting the maximum obtainable power from the generation sources [9, 10].

Power converters can also be controlled as *network-forming*, working equivalently as voltage sources, which is particularly applied in the islanded operation mode to manage the power requirements. However, that can also be implemented in the grid-connected mode, if necessary [8]. Network-forming converters support the main grid to determine the voltage components required to drive the non-dispatchable sources, by supplying the network feeders following power quality standards. Network-forming converters are synchronized to the grid using phase locked loop (PLL) functions, and are generally employed in conventional dispatchable generation units and energy storage devices [5, 7, 9]. Network-forming converters control the voltage amplitude and frequency at the PCC, producing as by-products active and reactive power signals [11].

1.2.2 Islanded Mode

Microgrid islanding operation mode can be intentional or unintentional. On the one hand, the intentional islanding may occur in scheduled maintenance cases or when the network power quality levels may jeopardize the microgrid operation. On the other hand, unintentional islanding occurs due to faults, contingencies, or other non-scheduled events. Islanded operation mode allows the continuity of supply (at least for the loads with higher priority), which represents cost savings and reliability improvements [3].

In this mode of operation, at least one converter must function as voltage source to dictate the network voltage conditions and power quality of the microgrid, which are followed by network-feeding converters. When there are more than one converter operating as a network-forming in autonomous operation, they must be synchronized using PLL function and should provide collaboratively the remaining power required by the loads [5, 9].

1.2.3   Transient Operating Mode

A transient operating mode refers to a state of transition between a grid-connected arrangement to an islanded mode, or vice versa. It carries additional challenges in electricity supply stability and grid protection and can occur due to intentional or unintentional reasons, such as grid constraints, predictive maintenance or safety issues [12, 13].

In microgrids, the disconnection and reconnection at the PCC must be seamlessly and as fast as possible, as it is crucial for the restoring process to be reliable with minimal or non-disturbance [7]. During the transient operating mode, it is necessary a synchronization procedure of voltage amplitude, frequency and phase, as well as coordinated sequence connections of distributed generation units, guaranteeing smooth transition [5].

To operate in both modes, grid-connected and islanded, the microgrid can be formed only by network-forming units, or cooperatively by network-forming and network-feeding units, in which the latter provide their maximum power and the former provide the residual power, ensuring a stable grid while verifying the utility grid conditions for restoration [2, 10].

1.3   Control Methods of Microgrids

This section describes microgrid control layers based on the hierarchical control method: primary, secondary and tertiary. The base layer controls the device-level and provides the fastest response, while the higher layers control the system-level with a slower response [2]. In order to guarantee power quality and disturbance rejection in microgrids, the essential functionalities of multilayer control techniques are regulating and providing stable voltage amplitude and frequency, through control loops that adjust active and reactive power flow, besides add filtering, harmonics current sharing and reactive power compensation capabilities [5]. To support the parallel operation of multiple voltage sources sharing the network loads and maintaining the power quality, the droop method is predominantly implemented on hierarchical controls [9]. The primary layer guarantees a precise power sharing between the inverters, by adding virtual inertia that emulates the physical characteristics of conventional power systems through voltage amplitude and frequency regulation. The secondary control restores the voltage amplitude and frequency deviations caused by the droop control. The tertiary layer manages the power flow among the microgrid and the utility grid at the PCC [10, 13, 14].

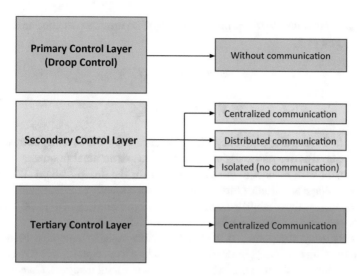

Fig. 1.1 Hierarchical control—layer approaches [3]

The primary layer is employed at each inverter based only on local measurements, while the secondary control can be implemented similarly localized or based on communication between the inverters, and in this last scenario the secondary control is arranged either centralized or distributed. On the other hand, the tertiary control layer must be centrally controlled with communication means [3]. The main aspects from each hierarchy layer of control are illustrated in Fig. 1.1.

1.3.1 Primary Control Layer

The primary control layer derived from the droop control method is implemented in order to manage the power supplied by each converter through voltage frequency and amplitude regulation. The droop control concept originates from high-power system, which permits large synchronous generators with high inertias to operate in parallel sharing the load by reducing their frequency when active power increases on the grid [2, 5, 10].

Contrary to typical conventional power systems, generation units electronically coupled to power converters do not incorporate inertia properties to provide stability on the system during synchronization stages. Instead it provides a fully control of the system dynamics and transient by offering a very fast response. Therefore, to enhance the microgrid stability and coordination of voltage sources operating in parallel, the inertia property of synchronous generators is electronically emulated in the network-forming converters by the droop method, regulating the voltage amplitude and frequency proportionally to the active and reactive power components [2, 5, 8, 10].

The basic equations for the primary layer based on the droop method are expressed as in (1.1) and (1.2):

$$\omega_i = \omega_{nom} - m_i P_i \tag{1.1}$$

$$V_i = V_{nom} - n_i Q_i \tag{1.2}$$

where ω_i is the angular frequency of the inverter i ($i = 1, 2, \dots n$) corresponding to the measured active power P_i, ω_{nom} is the network nominal frequency, and m_i is the coefficient with respect to the active power for the droop method. While, V_i is the output voltage amplitude of the converter proportionally to the measured reactive power Q_i, V_{nom} is the established nominal voltage amplitude, and n_i is the coefficient related to the reactive power for the droop method [3].

The frequency component is a global variable equally generated between the network-forming converters in steady-state, leading to an equal active power sharing when the droop coefficient m_i is constant and equal for all inverters. Conversely, the voltage component is a local variable, thus, even assuming similar droop coefficient n_i, there is no perfect reactive power sharing due to the uneven voltage amplitude in different nodes of the microgrid [3]. To illustrate this, for two network-forming converters the active and reactive power sharing relations are expressed in (1.3) and (1.4):

$$m_1 P_1 = m_2 P_1 \tag{1.3}$$

$$n_1 Q_1 = n_2 Q_1 + V_2 - V_1 \tag{1.4}$$

When the hierarchical control is not implemented, each converter injects dissimilar active power to the grid, depending on the line impedance detected at their nodes [14].

The angular frequency and voltage amplitude from the droop equations are used to generate the sinusoidal voltage reference of each converter, as described in (1.5):

$$V_{ref} = V_i \sin(\omega_i t) \tag{1.5}$$

In order to improve a dynamic transient response in the reference voltage, it is introduced a feed-forward component with respect to the active power in the angular frequency droop equation [3], as expressed in (1.6):

$$\omega_i = \omega_{nom} - m_i P_i - m_{ip} \frac{dP}{dt} \tag{1.6}$$

This droop compensation component corresponds to a proportional derivative (PD) controller, where the gain m_{ip} contributes to a faster transient response regarding active power variations, and the derivative term corresponds to the feed-forward signal [3]. This parameter can improve the position of the closed-loop poles of the system, providing a higher variability and reducing damping characteristics.

For the purpose of attenuating the distortion, reducing the impact of circulating current and ensuring harmonic current sharing under non-linear and unbalanced loads, the droop method emulates an impedance at the converter output through an additional closed-loop control, defined as virtual impedance [10, 13, 15, 16]. The virtual impedance is included in the voltage reference signal as a new variable based on the output current, as expressed in (1.7) [3, 10, 16]:

$$V_{ref} = V_i \sin(ph) - (R_v i_o + L_v \frac{di_o}{dt})$$

(1.7)

where ph corresponds to the integral over time of (1.6), i_o is the converter output current, and R_v and L_v are the resistive and reactive inductive components of the virtual impedance Z_v, expressed as in (1.8) [16]:

$$Z_v = R_v + jL_v$$

(1.8)

The virtual impedance variable is modified in order to provide mainly an inductive network, ensuring controllability of the active and reactive power by the droop Eqs. (1.1) and (1.2) [3].

Figure 1.2 illustrates the virtual output impedance loop regarding the droop control, voltage and inner current loops [10]. The virtual output impedance can provide further functions, as soft-start operation provided by designing a higher value of impedance at the start, and then gradually decreasing it [5, 10, 13, 15].

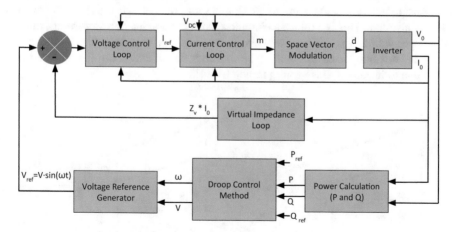

Fig. 1.2 Droop control with virtual output impedance control loop [10]

The droop control method applied in the primary control offers a satisfactory power balance among converters designed as network-forming. However, it introduces an error in the steady-state frequency and voltage amplitude, which is addressed by the secondary control [5].

1.3.2 Secondary Control Layer

The secondary control layer mitigates the deviations in frequency and voltage amplitude caused by the droop control in the steady-state, restoring their values to specific references and maintaining the power sharing accomplished by the primary layer. In this layer, the restoration of the frequency and voltage to nominal values is given by adding a corrective term and shifting up the droop function to the initial characteristics of each unit [5, 14]. Figure 1.3 illustrates the primary and secondary control actions.

It is notable that, without a secondary layer implementation, both frequency and voltage amplitude of generation in the microgrid are load dependent, with deviations originating on the virtual inertia and impedances of the droop control in the primary layer [10, 13].

There are basically four secondary control techniques, varying from one-to-all (central control) or all-to-all (distributed control) traffic schemes of communication services [9, 14, 17].

(1) *Centralized Control*: is based on a central controller that requires one to all communication scheme. The central controller computes all components errors, using measurements from the PCC, and sends the corrective term to the other converters. This is a robust technique against communication constraints, however with low reliability (network is completely central dependant), tolerance and flexibility in case of fault of the converter, demanding controller duplication due to its master-slave configuration [9, 14].

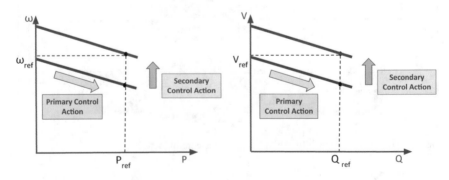

Fig. 1.3 Primary and secondary control actions [10]

(2) *Decentralized Control*: is a hybrid system based on local and central controllers, that requires one to all communication schemes. As in the former technique, the signals errors are common to the inverters obtained by the central controller, but further properties as the corrective term are autonomously computed by each local controller [14]. This technique is sensitive to communication issues, since the inverters can produce different corrective terms if receiving information in distinct time due to non-synchronized internal clocks between controllers, an effect known as clock drift [9].

(3) *Averaging Control*: is based on a decentralized controller that requires all-to-all communication scheme. Each controller calculates the error by averaging the output values from all other controllers. This technique also presents limitations due to communication performance between the controllers, incurring in incorrect power sharing in case of data losses [14].

(4) *Consensus Control*: is based on a decentralized control that requires all-to-all communication scheme. Each controller computes its own corrective term based on averaging the components deviation among the local and neighbour inverters, the local components errors, and the averaging of the corrective terms error among the inverters. The decision-making agents in the network communicate in a peer-to-peer protocol with its neighbouring nodes, aiming for a global consensus on power control [2]. This strategy guarantees the power sharing and proper frequency and voltage amplitude regardless of communication constraints. However, it requires superior computing power to process an increased data exchange rate, which impacts the capital costs of the system [14].

The main aspects of each secondary control technique are qualitatively summarized in Table 1.1, which evaluates the type of service scheme for communication, the reliability and tolerance of the system in case of fault of the converters, if clock drift presents a significant impact due to different time signals from each controller, the communication constraints regarding loss of packages and the required communication bandwidth concerning the network data exchange rate.

Table 1.1 Main aspects of each secondary control technique

Technique	Communication scheme	Fault tolerance	Clock drift	Communication constraints	Communication bandwidth
Centralized control	One-to-all	Weak	No	Robust	High
Decentralized local control	One-to-all	Strong	Yes	None	None
Decentralized averaging control	All-to-all	Strong	No	Weak	Very high
Decentralized consensus control	All-to-all	Strong	No	Strong	Very high

Additional to the secondary control methods based on communication services, there are some proposed techniques that do not require communication means, such as the proposed by [18].

1.3.3   Tertiary Control Layer

In the tertiary control layer, the global power flow between the microgrid and the main grid is optimized by autonomously sensing unbalances in the energy supply and demand [8]. The control of the energy imported and exported from the utility grid depends on technical and economic issues, such as harmonics mitigation, grid faults tolerance due to fluctuation, imbalance or interruption, and dynamic price of energy, mainly in peak periods [5]. A strong implementation of the tertiary control layer is thus vital for creating an efficient bidirectional connection to control the power flow [13].

The tertiary control layer, sometimes referred broadly as energy management system (EMS) or grid management system within the industry, is applied in the traditional centralized grid configuration to optimize operation costs with real time monitoring and automated decision-making based on generation costs, weather forecasts and market information. These functionalities, as well as additional ones such as real and reactive power support, intentional islanding and ancillary services can be adapted and implemented in a microgrid scenario, in what is described as a microgrid management system (MGMS), communicating with inferior control layers for centralized control of the system's power flow when operating in grid-connected mode [2, 9].

In the grid-connected mode, the power flow can be controlled by regulating the voltage amplitude and frequency, as in the operation of the primary and secondary control from the islanded mode. Nevertheless, after proper synchronization with the main grid, the power flow bidirectionality can be controlled by the tertiary layer, which defines the desired active and reactive power values by sending the voltage amplitude and frequency references to the secondary control. To avoid instabilities in the network, the tertiary control references should be disabled in case of islanding detection, when the utility is disconnected from the microgrid [5, 10].

1.4   Conclusion

This chapter introduced an overview of the future of the electrical power grid, from centralized, vertical and unidirectional policies and strategies to distributed, multi-level and bidirectional flow systems. Microgrids, a local controllable electrical power system that integrates generation, storage and consumption, and can operate in grid-connected, islanded and transient operating mode, play a significant part in enabling this restructuring. Connected to the main grid, the microgrid aims to support and enhance the network stability and reliability, while disconnected (islanded) it must sustain the required power quality of the grid by itself. The transition between both grid-connected and islanded modes needs to be smooth during disconnection and restoration operations. In this chapter the droop method, largely implemented on microgrid hierarchical controls, was described in primary, secondary and tertiary layers' control. Implemented in power inverters, these control strategies adjust active and reactive power flow in order to provide stable voltage amplitude and frequency and guarantee the network's power quality. The primary control layer ensures power sharing between the inverters with the droop control method, by emulating the physical characteristics of traditional power systems through a virtual inertia factor that regulates voltage amplitude and frequency components. The droop control method from the primary layer origin deviations on the voltage amplitude and frequency components that is improved by the secondary control layer. The tertiary control layer coordinates the optimal power flow between the microgrid and the main grid. The correct implementation of microgrids and their control structures are vital to fully reap its benefits in terms of operating cost reduction, increased grid stability and performance, and environmental impacts mitigation.

References

1. El-Hawary ME (2008) Introduction to electrical power systems, vol 50. Wiley, New Jersey
2. Cheng Z, Duan J, Chow M-Y (2018) To centralize or to distribute: that is the question: a comparison of advanced microgrid management systems. IEEE Ind Electron Mag 12(1):6–24
3. de Souza ACZ, Castilla M (2019) Microgrids design and implementation, 1st edn. Springer, Cham
4. Hatziargyriou N, Asano H, Iravani R, Marnay C (2007) Microgrids. Power Energy Mag IEEE 5(4):78–94
5. Vasquez JC, Guerrero JM, Miret J, Castilla M, De Vicuna LG (2010) Hierarchical control of intelligent microgrids. IEEE Ind Electron Mag 4(4):23–29
6. Moslehi K, Kumar R (2010) A reliability perspective of the smart grid. IEEE Trans Smart Grid 1(1):57–64
7. Rocabert J, Azevedo GM, Luna A, Guerrero JM, Candela JI, Rodríguez P (2011) Intelligent connection agent for three-phase grid-connected microgrids. IEEE Trans Power Electron 26(10):2993–3005
8. Guerrero JM, Loh PC, Lee T-L, Chandorkar M (2012) Advanced control architectures for intelligent microgrids—Part II: Power quality, energy storage, and AC/DC microgrids. IEEE Trans Industr Electron 60(4):1263–1270
9. Miret J, García de Vicuña J, Guzmán R, Camacho A, Moradi Ghahderijani M (2017) A flexible experimental laboratory for distributed generation networks based on power inverters. Energies 10(10):1589
10. Guerrero JM, Chandorkar M, Lee T-L, Loh PC (2012) Advanced control architectures for intelligent microgrids—Part I: Decentralized and hierarchical control. IEEE Trans Industr Electron 60(4):1254–1262
11. Yazdani A, Iravani R (2010) Voltage-sourced converters in power systems, vol 34. Wiley Online Library, New Jersey
12. Chitra N, Sivakumar P, Priyanaka S, Devisree A (2018) Transient behaviour of a microgrid and its impact on stability during pre and post islanding—a novel survey. Int J Eng Technol 7(2.24):326–330

13. Guerrero JM, Vasquez JC, Matas J, De Vicuña LG, Castilla M (2010) Hierarchical control of droop-controlled AC and DC microgrids—a general approach toward standardization. IEEE Trans Industr Electron 58(1):158–172

14. Martí P, Velasco M, Martín EX, de Vicuña LG, Miret J, Castilla M (2016) Performance evaluation of secondary control policies with respect to digital communications properties in inverter-based islanded microgrids. IEEE Trans Smart Grid 9(3):2192–2202

15. Guerrero JM, Matas J, de Vicuña L (2006) Wireless-control strategy for parallel operation of distributed generation inverters. IEEE Trans Ind Electron 53(5):1461–1470

16. Matas J, Castilla M, De Vicuña LG, Miret J, Vasquez JC (2010) Virtual impedance loop for droop-controlled single-phase parallel inverters using a second-order general-integrator scheme. IEEE Trans Power Electron 25(12):2993–3002

17. Guo F, Wen C, Mao J, Song Y-D (2014) Distributed secondary voltage and frequency restoration control of droop-controlled inverter-based microgrids. IEEE Trans Industr Electron 62(7):4355–4364

18. Rey JM, Martí P, Velasco M, Miret J, Castilla M (2017) Secondary switched control with no communications for islanded microgrids. IEEE Trans Industr Electron 64(11):8534–8545

Chapter 2
Case Studies: Modelling and Simulation

Abstract Scientific research today is focused on creating and optimizing algorithms and hardware that improve the controlling techniques of microgrids, making their adoption viable and increasingly advantageous. In order to evaluate the efficacy of these novel approaches, computer simulations are often developed. In the academic environment it is well known that the integrated software MATLAB-Simulink is a powerful, robust and versatile tool able to perform those tasks. This chapter provides a detailed guideline for design and simulation on the MATLAB-Simulink software platform of basic control methods applied to microgrids on different operating modes, with discussions on performance for each configuration. In the grid-connected operating mode, a system of one network-feeding converter and one local load is studied. In the islanded mode, it is evaluated network-forming converters with local and common load connected to the grid. Finally, in the transient operating mode, a study is proposed of a grid with one network-feeding and one network-forming converter, as well as one common load, in order to investigate features from disconnection and re-connection procedures between the two grid-connected and islanded mode.

Keywords Microgrid modelling · Microgrid simulation · Droop control · Matlab-Simulink simulation · Network-feeding converter · Network-forming converter

Electronic supplementary material The online version of this chapter (https://doi.org/10.1007/978-3-030-43013-9_2) contains supplementary material, which is available to authorized users.

2.1 Case Study I: Grid Connected Mode—Microgrid Composed by One Network-Feeding Converter and One Local Load

2.1.1 Function and Control Parameters

The Case Study I presents a grid-connected operation mode with one inverter operating as a current source connected in parallel with one local load. In this operating mode, the utility grid is defining the voltage amplitude, frequency and phase. The main objective of this case, therefore, is to control the active and reactive power imported and exported from and to the main grid, by adjusting the modulating signal produced in the inner current loop.

The electrical parameters adjusted for the microgrid to be studied are presented in the Table 2.1. The defined values for the nominal and DC-link voltage are based on the physical characteristics of the buck converter with step-down behaviour. A typical range for the voltage values are detailed in the Table 2.2.

The nominal AC_{RMS} voltage, load resistance and nominal frequency components were determined in order to limit the load inductance in a per unit value between

Table 2.1 Microgrid electrical parameters

Parameter name		Acronym	Value	Units
	Nominal AC voltage	V_{nom}	230	V
	Nominal frequency	f_{nom}	50	Hz
	Nominal angular frequency	ω_{nom}	$2\pi \times f_{nom}$	rad/s
	DC-link voltage	V_{DC}	800	V
	Load resistance	R_L	22	Ω
	Load inductance	L_L	5	mH
	Grid resistance	R_g	65	mΩ
	Grid inductance	L_g	1	mH
Inverter 1 l	Output inductance	L_{o1}	5	mH
Inverter 1 l	Output resistance	R_{o1}	0.5	Ω
Inverter 1 l	LC filter capacitance	C_{f1}	10	μF
Inverter 1 l	LC filter damping resistor	R_{f1}	20	Ω
Inverter 1 l	Line inductance	L_{L1}	1	mH
Inverter 1 l	Line resistance	R_{L1}	65	mΩ

Table 2.2 Characteristic range for DC-link and AC voltage

DC-link voltage (V)	AC voltage (V)
350–650	110
650–850	230

0.05 and 0.10, as well as to achieve a desired output power range. The equations adopted for defining these parameters are described as in (2.1):

$$P = 3 \times \frac{V_{AC}^2}{R_L} \tag{2.1}$$

From (2.1), the active power output is nearly 7 kW, which leads to a current of about 10 A, according to (2.2):

$$I = \frac{P}{3 \times V_{AC}} \tag{2.2}$$

Considering the approximation stated by (2.3), for L_{o1} equal to 5 mH, the per-unit value calculated is between the acceptable range, corresponding to 0.07:

$$Z = \frac{V_{AC}}{I} \cong R_L \Rightarrow \frac{(\omega_{nom} \times L_{o1})}{Z} = 0.07 \tag{2.3}$$

The resistance and inductance load values determine the load to be more resistive than inductive.

The values for impedance on the output of the converter, local and grid line characterize the effects of losses and parasitic elements of the power cables and network configuration [1]. The defined values detailed in the Table 2.1 were based on the parameters studied by [2] and configured by the experimental laboratory for distribution generation networks at UPC, as presented in [3].

The converter output interfaces with the load through a filter. The shunt capacitor and damping resistor values for the filter were fine-tuned in order to reduce noise and enhance the power output during disturbances in the load current. The low impedance of the filter capacitor provides a bypass for harmonics by switching the harmonic currents from the converter, avoiding them to enter into the load, also guaranteeing voltage support at the node [4, 5].

The control parameters applied to the Case Study I are presented in the Table 2.3.

The sampling rate in the microsecond order increases the running time for the simulation, however it was chosen since it is a reliable value to evaluate the control methods and parameters selected for all the case studies.

The cut-off frequency value is added on a low pass filter (LPF), which is implemented to attenuate both harmonics and noise in the output voltage and power signals. The selected cut-off frequency follows the rules defined by [6], which states that this frequency should be between one to two decades lower than the nominal frequency, in order to ensure noise rejection and slow transient response, one desirable characteristic in electrical power systems.

The design of the control parameters for the compensators can vary basically with different operation points, sampling rate and power range. For all the case studies, the established control values are equivalent and were tuned while performing the simulations based on the core functions and characteristics of each parameter.

Table 2.3 Microgrid control parameters

Parameter name	Acronym	Value	Units
Sampling rate	t_s	1	μs
Cut-off frequency LPF	ω_c	$0.03 \times \omega_{nom}$	rad/s
Proportional gain PI dynamic active power reference	k_{pp}	0	A^{-1}
Integral gain PI dynamic active power reference	k_{ip}	0.5	$(As)^{-1}$
Proportional gain PI dynamic reactive power reference	k_{pq}	0	A^{-1}
Integral gain PI dynamic reactive power reference	k_{iq}	0.5	$(As)^{-1}$
Proportional gain PRES current compensator	k_{pi}	12	A^{-1}
Integral gain PRES current compensator	k_{ii}	200	$(As)^{-1}$
Damping coefficient PRES current compensator	shi_i	0.1	

2.1.2 Scheme and Subsystems

For the purpose of this case study, the model designed presents a DC source that emulates the distributed generator, three-phase half-bridge insulated-gate bipolar transistor (IGBT) switches from S1 to S6—representing a DC to AC converter—operated by controlling the modulating signal, as well as electrical components, AC power grid emulator and control subsystem.

The scheme of the microgrid composed by one network-feeding converter modelled in series with one local load and the grid is presented in Fig. 2.1.

Figure 2.2 presents the inverter control subsystem composed by three blocks: Alpha_Beta which is responsible for converting a three dimensional phasor into a two dimensional orthogonal phasor frame—defined as alpha and beta ($\alpha\beta$); the combined Power and Current Control Loop, which generate the modulating signal in respect with a pre-defined reference value for active and reactive power; and Space Vector Modulation, which transforms the modulating signal given in $\alpha\beta$ axis components to pulses for each of the six switches from the three half-bridge converters.

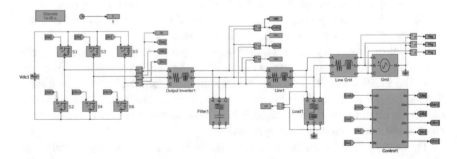

Fig. 2.1 Scheme of the microgrid composed by one network-feeding inverter, one local load and the power grid

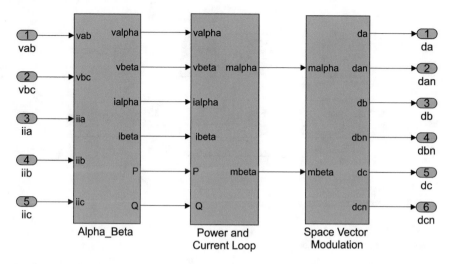

Fig. 2.2 Scheme of the control subsystem

Aiming to simplify the analysis and control of three-phase converters, the three-dimensional phasor terms are represented by means of αβ-frame components, expressed as in (2.4) by the simplified transform of balanced systems, defined by the Clarke Transform [5, 7]:

$$\begin{bmatrix} f_\alpha(t) \\ f_\beta(t) \end{bmatrix} = \frac{2}{3} \begin{bmatrix} 1 & -\frac{1}{2} & -\frac{1}{2} \\ 0 & \frac{\sqrt{3}}{2} & -\frac{\sqrt{3}}{2} \end{bmatrix} \begin{bmatrix} f_a(t) \\ f_b(t) \\ f_c(t) \end{bmatrix} \tag{2.4}$$

The amplitude and sinusoidal functions of the αβ reference frame are expressed as in (2.5), (2.6), and (2.7) [5]:

$$\hat{f}(t) = \sqrt{f_\alpha^2 + f_\beta^2} \tag{2.5}$$

$$f_\alpha(t) = \hat{f}(t) \cos[\theta(t)] \tag{2.6}$$

$$f_\beta(t) = \hat{f}(t) \sin[\theta(t)] \tag{2.7}$$

In the simulation, the three-dimensional phase is converted to αβ-frame using Clarke Transform, as illustrated in Fig. 2.3.

The integration between MATLAB and Simulink allows incorporating algorithms into the models. Thus, the MATLAB code implemented for the Clarke Transform subsystem is described as follows:

```
function [alpha,beta] = fcn2(ab,bc)
alpha =(ab*2+bc)/3;
```

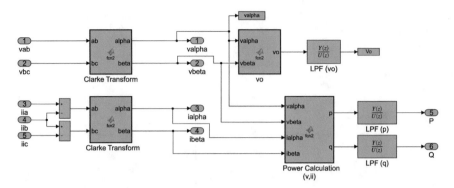

Fig. 2.3 Scheme of the αβ-frame subsystem

```
beta =(sqrt(3)/3)*bc;
```

This transform is applied to generate the output voltage resulting in *valpha, vbeta*, which are variables used to determine the amplitude of the output voltage value v_o as in (2.5), described by the following code:

```
function vo = fcn2(valpha,vbeta)
vo = sqrt(valpha^2+vbeta^2);
```

The αβ-frame is also implemented to convert the current components, defined as *ialpha* and *ibeta*, which combining with the *valpha* and *vbeta* are variables responsible for controlling the active and reactive power signals, as proposed by [3, 5], with the expressions (2.8) and (2.9):

$$p = \frac{3}{2}\left(v_\alpha i_\alpha + v_\beta i_\beta\right) \tag{2.8}$$

$$q = \frac{3}{2}\left(-v_\alpha i_\beta + v_\beta i_\alpha\right) \tag{2.9}$$

The active and reactive power components in αβ-frame algorithm applied in the MATLAB function is described as follows:

```
function [p,q] = fcn2(valpha,vbeta,ialpha,ibeta)

p = (3/2)*(valpha*ialpha+vbeta*ibeta);
q = (3/2)*(-valpha*ibeta+vbeta*ialpha);
```

Both output voltage and power components include LPF for noise and harmonic mitigation, as well as for slow dynamic response. The transfer function in the Laplace domain for each element is represented by (2.10), (2.11) and (2.12) [3, 6]:

$$V_o(s) = v_o(s)\frac{w_c}{s + w_c} \tag{2.10}$$

$$P(s) = p(s)\frac{w_c}{s + w_c} \tag{2.11}$$

$$Q(s) = q(s)\frac{w_c}{s + w_c} \tag{2.12}$$

In the MATLAB algorithm, all transfer functions are converted from continuous to discrete-time dynamic system model at the sample rate t_s, using Tustin integration, which is based on the Backward Euler method with suitable precision and numerical stability.

Based on the converter output voltage in $\alpha\beta$ frame and the established reference values in unit of [W] for active and [VAr] for reactive power, the power control loop produces a reference current. These currents components, defined as *ialpharef* and *ibetaref*, are implemented in an inner current control loop to produce modulating signals for the IGBT bridge switches [3]. Figure 2.4 illustrates the power and current control loops along with a dynamic power reference block.

The dynamic power reference block aims to guarantee the defined reference active and reactive power values at the steady-state response, by comparing the reference values with the estimated from the current and voltage measurements and mitigating the error through a proportional integral (PI) compensator. The scheme of the dynamic power reference subsystem is illustrated in Fig. 2.5.

The PI transfer functions introduced in the dynamic power reference subsystem are following described by (2.13) and (2.14).

$$PI(s) = \frac{k_{pp}s + k_{ip}}{s} \tag{2.13}$$

$$PI(s) = \frac{k_{pq}s + k_{iq}}{s} \tag{2.14}$$

The simulation code implemented for the Power Control Loop subsystem is presented as follows:

```
function [ialpharef,ibetaref] = fcn2(valpha,vbeta,pref,qref)
```

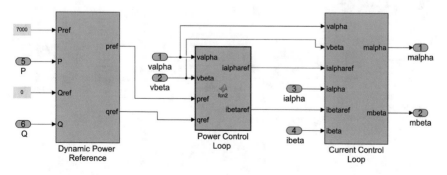

Fig. 2.4 Scheme of the power and current control loop subsystem

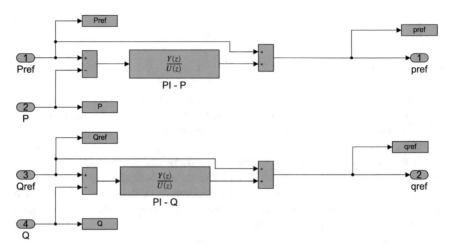

Fig. 2.5 Scheme of the dynamic power reference subsystem

```
den = 1/(valpha^2+vbeta^2);
ialpharef = (2/3)*(valpha*pref+vbeta*qref)*den;
ibetaref = (2/3)*(vbeta*pref-valpha*qref)*den;
```

In the inner Current Control Loop subsystem, one proportional plus resonant compensator (PRES) is applied to process and eliminate the current error generated from the comparison between the reference current and the output current, producing an error control signal, as illustrated in Fig. 2.6. In this loop a feed-forward term, corresponding to the output voltage, is applied to mitigate the dynamic coupling

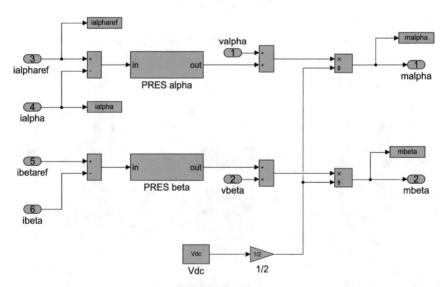

Fig. 2.6 Scheme of the current control loop subsystem

between the converter output and the network by accelerating the response against external disturbances. The feed-forward term, added to the control signal is multiplied by $V_{dc}/2$, due to the supplied split voltage of half-bridge converter, generating the modulating signals that drives space vector modulation pulses for the converters switches [3].

Figure 2.7 illustrates the PRES subsystem, while (2.15) and (2.16) describe the PRES transfer functions. The compensator proportional gain k_{pi} was tuned to provide an effective transient response, while the integral gain k_{ii} and the damping coefficient *shi* were designed to mitigate the current error and enhance the steady-state response.

$$P_k(s) = k_{pi} \tag{2.15}$$

$$RES(s) = \frac{k_{ii}2\,shi_i\,\omega_{nom}s}{s^2 + 2shi_i\omega_{nom}s + \omega_{nom}^2} \tag{2.16}$$

By using space vector modulation (SVM), the modulating signals, called as *malpha* and *mbeta*, generated by the current control loop, are converted to pulses to each of the half-bridge switches, as illustrated in Fig. 2.8. However, in order to agree with

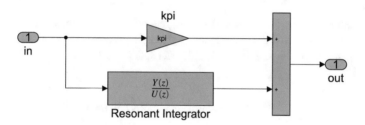

Fig. 2.7 Scheme of the PRES subsystem

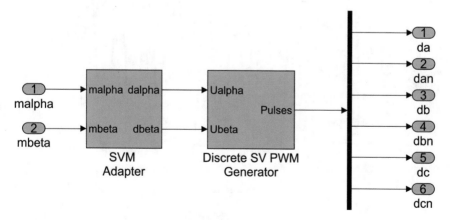

Fig. 2.8 Scheme of the SVM block

Fig. 2.9 Scheme of SVM
adapter subsystem

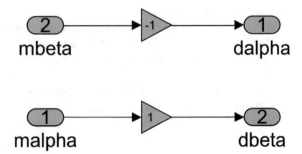

the properties and characterization of the subsystem regarding the discrete space
vector pulse width modulation generator (SV PWM) implemented in this study, proposed by [8], an adaptation from the modulating alpha and beta components had to
be applied, as demonstrated in Fig. 2.9.

2.1.3   Simulation Results

The simulation results for the modulating signal, current tracking, as well as the output voltage response and the instantaneous grid voltages, and the power components
are demonstrated in Figs. 2.10, 2.11, 2.12, 2.13, 2.14, 2.15, 2.16, 2.17, 2.18, 2.19,
2.20 and 2.21.

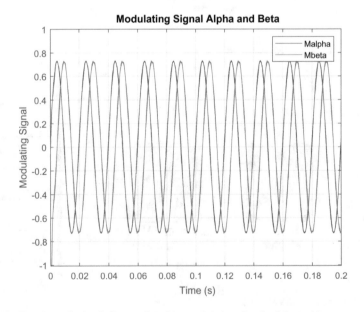

Fig. 2.10 Case Study I: simulation results of the modulating signals alpha and beta

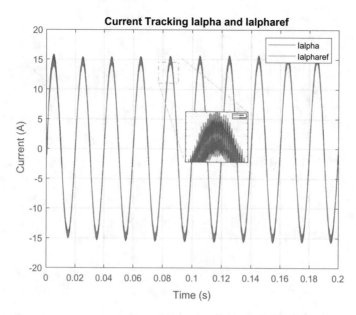

Fig. 2.11 Case Study I: simulation results of the current tracking from current control loop with zoom

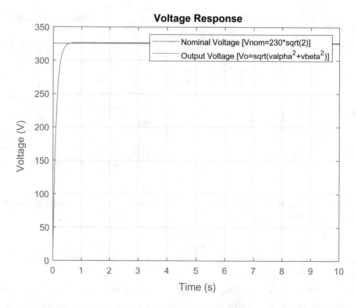

Fig. 2.12 Case Study I: simulation results of the nominal and output voltage

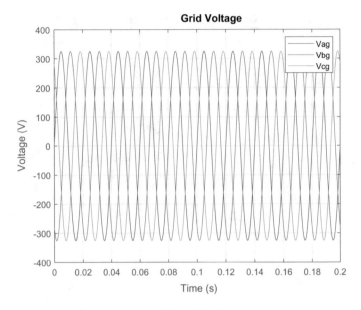

Fig. 2.13 Case Study I: simulation results of the grid voltage

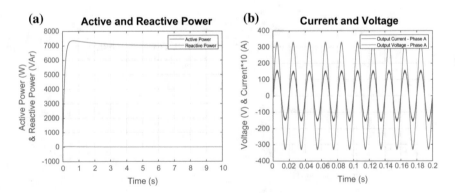

Fig. 2.14 Case Study I: simulation results of 7 kW and 0 kVAr, **a** active and reactive power curves, **b** phase between the voltage and the current

The modulating signal from Fig. 2.10 provides an accurate response by presenting the alpha leading the beta component by 90° ($\pi/2$).

Figure 2.11 proves the effective operation from the current control loop, in which the alpha current component follows the reference current established by the power control loop based on the output voltage measured values and the defined active and reactive power.

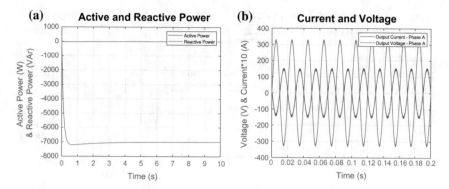

Fig. 2.15 Case Study I: simulation results of −7 kW and 0 kVAr, **a** active and reactive power curves, **b** phase between the voltage and the current

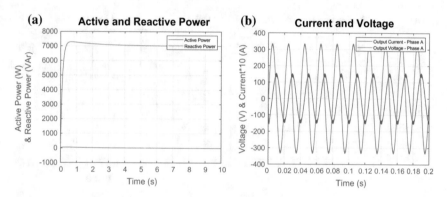

Fig. 2.16 Case Study I: simulation results of 0 kW and 7 kVAr, **a** active and reactive power curves, **b** phase between the voltage and the current

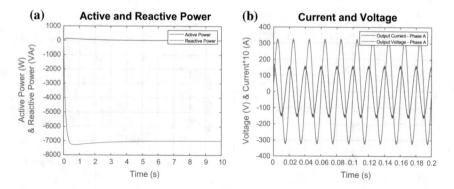

Fig. 2.17 Case Study I: simulation results of 0 kW and −7 kVAr, **a** active and reactive power curves, **b** phase between the voltage and the current

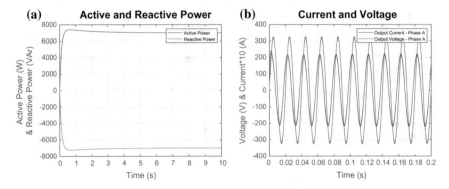

Fig. 2.18 Case Study I: simulation results of 7 kW and −7 kVAr, **a** active and reactive power curves, **b** phase between the voltage and the current

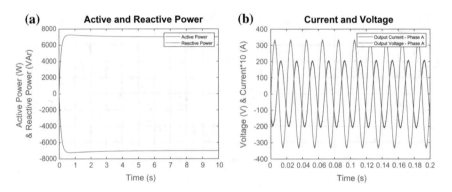

Fig. 2.19 Case Study I: simulation results of −7 kW and 7 kVAr, **a** active and reactive power curves, **b** phase between the voltage and the current

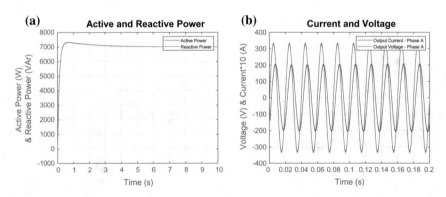

Fig. 2.20 Case Study I: simulation results of 7 kW and 7 kVAr, **a** active and reactive power curves, **b** phase between the voltage and the current

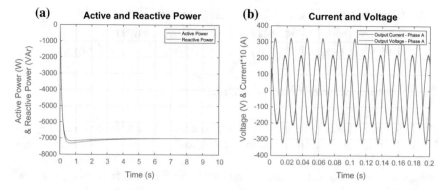

Fig. 2.21 Case Study I: simulation results of −7 kW and −7 kVAr, **a** active and reactive power curves, **b** phase between the voltage and the current

The result presented by Fig. 2.12 demonstrates the stable operation of the system, which achieves a satisfactory signal performance for the voltage response, reaching its nominal value.

The grid voltages illustrated in Fig. 2.13 presents a perfect balanced sinusoidal system with constant frequency, which is imposed to the microgrid.

Figs. 2.14, 2.15, 2.16, 2.17, 2.18, 2.19, 2.20 and 2.21 present the curves of active and reactive power, as well as the current and voltage generated in phase A for different active and reactive power values.

Table 2.4 summarizes the defined active and reactive power components in the model and the response of phase between the voltage and the current for each scenario.

Table 2.4 Case Study I: active and reactive power components, and current and voltage in phase A

Test	Active power (kW)	Reactive power (kVAr)	Current in respect to voltage
1	7	0	In phase
2	−7	0	Lagging by π
3	0	7	Lagging by $\pi/2$
4	0	−7	Leading by $\pi/2$
5	7	−7	Leading by ~$\pi/4$
6	−7	7	Lagging by ~$3\pi/4$
7	7	7	Lagging by ~$\pi/4$
8	−7	−7	Leading by ~$3\pi/4$

2.2 Case Study II: Grid Connected Mode—Microgrid Composed by One Network-Feeding Converter and One Local Load with Additional Control Loops

2.2.1 Function and Control Parameters

The Case Study II presents a grid-connected operation mode with one inverter operating as a current source and one local load. In addition to the functionalities introduced in the Case Study I, it is implemented a DC-link voltage control to adjust the DC voltage to a reference value, which is applied in MPPT converters controls, as well as an output voltage control to regulate the AC voltage to a defined value.

In order to implement the DC-link voltage control, the DC source applied in the Case Study I to emulate the distributed generator needs to be replaced by one current source in parallel with one capacitor. This circuit represents the operation of the boost converter with a MPPT algorithm, which guarantees the maximum power production from the energy source by estimating a suitable value for the reference voltage V_{DC} [6]. The simulation defined values for both components, current source and capacitor at the DC-link, are presented in Table 2.5. For a fast-transient response, the capacitor is set to an initial voltage of 80% of the designed V_{DC} reference value.

Additionally to the control parameters from the Case Study I, Table 2.6 presents further parameters established for the Case Study II. The slope gain that is applied in the output voltage loop, representing the feedback term that is combined with an integrator compensator, is designed as in (2.17):

$$k_q = \frac{\Delta V}{Q_{max}} = \frac{2}{3} \frac{\omega_{nom} L_g}{V_{min}} \tag{2.17}$$

Table 2.5 Additional microgrid electrical parameters

Parameter name	Acronym	Value	Units
DC-link current	I_{DC}	4	A
DC-link capacitor	C_{DC}	1000	μF

Table 2.6 Additional microgrid control parameters

Parameter name	Acronym	Value	Units
Proportional gain PI DC-link voltage compensator	k_{pVdc}	30	A/V
Integral gain PI DC-link voltage compensator	k_{iVdc}	325	A/(Vs)
Proportional gain PI output voltage compensator	k_{poV}	0	A/V
Integral gain PI output voltage compensator	k_{ioV}	25,000	A/(Vs)
Slope gain output voltage compensator	k_q	1	mΩ/V

As stated before, the design of the others control parameters for the compensators were adjusted while modelling the system based on the core functions and characteristics of each parameter.

2.2.2   Scheme and Subsystems

For the purpose of this case, the model designed is similar to the Case Study I scheme from Fig. 2.1, replacing the DC voltage source to a DC controllable current source, and adding DC-link capacitor and voltage measurement units, as illustrated in Fig. 2.22.

In this case study, the control subsystem includes additional voltage control approaches, the DC-link and output voltage controls, as presented in Fig. 2.23. Note that in this case, P and Q are not specified, but generated from the DC-link and output voltage controls, respectively.

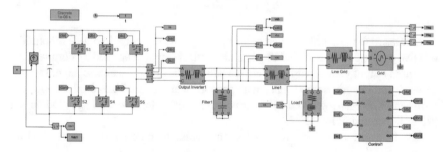

Fig. 2.22 Scheme of the microgrid composed by one network-feeding converter, one local load and the power grid with additional control loops

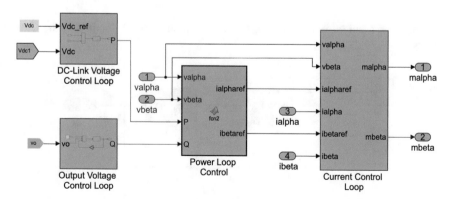

Fig. 2.23 Scheme of the power and current control subsystem

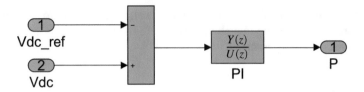

Fig. 2.24 Scheme of the DC-Link voltage control loop subsystem

In order to adjust the DC-link voltage, it is injected (or absorbed) active power to (or from) the network corresponding to the control signal produced by comparing the measured DC-link voltage and a reference value. In the control loop, a PI compensator process and eliminate the error generated by the voltage comparison, as illustrated in Fig. 2.24.

The PI transfer function from the DC-link voltage control loop is described by (2.18):

$$PI(s) = \frac{k_{pVdc}s + k_{iVdc}}{s} \tag{2.18}$$

Similar to the DC-link voltage control, to regulate the output voltage amplitude, reactive power flow is adjusted through the control signal generated by comparing a measured output voltage with a reference value. This produced control signal is processed by an integral compensator with proportional feedback based on a slope gain term, which offers a regulated dynamic response with a degraded voltage modulation [9], as illustrated in Fig. 2.25.

The PI transfer function from the output voltage control loop is described by (2.19):

$$PI(s) = \frac{k_{poV}s + k_{ioV}}{s} \tag{2.19}$$

The power and current control loops were implemented as detailed in the Case Study I.

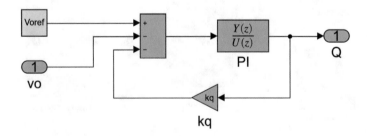

Fig. 2.25 Scheme of the output voltage control subsystem

2.2.3 Simulation Results

The simulation results for the modulating signal, current tracking, active and reactive power, as well as the output voltage response and instantaneous grid voltages are demonstrated in Figs. 2.26, 2.27, 2.28, 2.29 and 2.30.

As in the Case Study I, the modulating signal from Fig. 2.26 produces a precise response with the alpha leading the beta component by 90° ($\pi/2$).

Figure 2.27 presents an adequate operation of the current control loop by the alpha current component tracking the reference current established by the power control loop, which is based on the DC-link and output voltage control loops.

The current and voltage phase relation from Fig. 2.28 are the result of the signals produced for the active and reactive power due to the voltage control loops.

Figure 2.29 demonstrates the operation of both voltage loops implemented in this study. The output voltage control, which was applied intending to achieve 97% of the nominal voltage, in order to limit the amount of injected reactive power by the network-feeding converter. The DC-link voltage loop, which reaches the reference value defined in the simulation, modelling the response from the operation of the MPPT algorithm.

Figure 2.30 demonstrates the grid voltage represented by a stable three-phase sinusoidal system with a constant frequency, which is imposed to the microgrid.

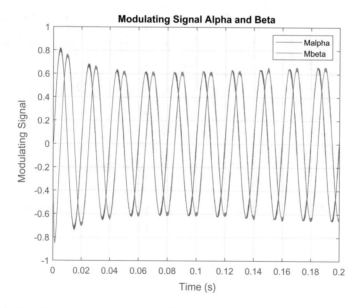

Fig. 2.26 Case Study II: simulation results of the modulating signal alpha and beta

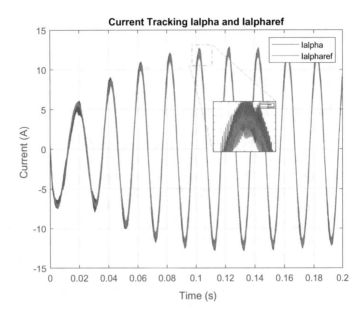

Fig. 2.27 Case Study II: simulation results of the current tracking from current control loop with zoom

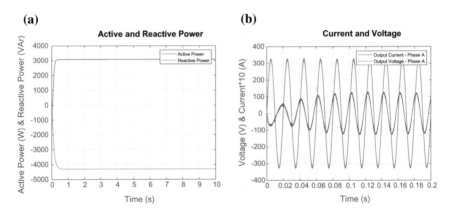

Fig. 2.28 Case Study II: simulation results of **a** active and reactive power, **b** current and voltage in phase A

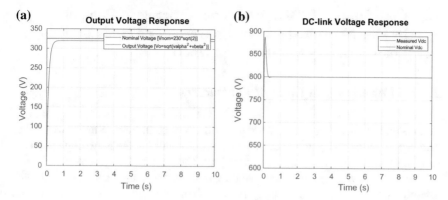

Fig. 2.29 Case Study II: simulation results of **a** output voltage loop, **b** DC-link voltage loop

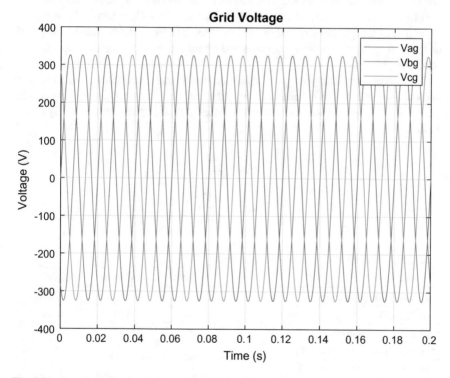

Fig. 2.30 Case Study II: simulation results of the grid voltage

2.3   Case Study III: Islanded Mode—Microgrid Composed by One Network-Forming Converter and One Local Load (Primary Control Loop Only)

2.3.1   Function and Control Parameters

In an islanded operating mode, the microgrid is not connected to the power utility grid due to pre or non-planned events, and in this condition the network feeders are supplied by network-forming converters [10]. Case Study III presents an off-grid operation mode with one inverter operating as a voltage source and one local load. Therefore, the purpose of this study is to implement an inner current loop and an outer voltage loop based on the droop control, which adjusts the frequency and the voltage magnitude with respect to active and reactive power signals, respectively.

Table 2.7 presents extra parameters established for the Case Study III. Similar to the compensator parameters, the definition of the droop control values can vary with different scenarios, and were also obtained by tuning each parameter to the desired performance.

The droop control slopes m_p and n_q are defined in order to optimize the frequency and voltage profile in reference with the active and reactive signal, respectively. The design of the compensator's gains can be based on the relation expressed by (2.20) and (2.21) [4]:

$$m_p = \frac{\Delta f}{P_{max}} \tag{2.20}$$

$$n_q = \frac{\Delta V}{2Q_{max}} \tag{2.21}$$

Table 2.7 Additional microgrid control parameters

Parameter name	Acronym	Value	Units
Frequency droop parameter	m_p	130	μrad/Ws
Frequency droop derivative parameter	m_{pp}	10	μrad/W
Voltage droop parameter	n_q	1	mV/VAr
Droop method virtual impedance	L_V	1	mH
Droop method virtual resistance	R_V	0	mH
Proportional gain PRES voltage compensator	k_{pv}	0.1	A/V
Integral gain PRES voltage compensator	k_{iv}	0.1	A/(Vs)
Damping coefficient PRES voltage compensator	shi_v	0.01	

The virtual output impedance design is proposed by [6], based on the statement that a dominant inductive characteristic can be satisfactorily ensured by any impedance angle higher than 65°, as expressed in (2.22) and (2.23):

$$\theta = \tan^{-1}\left(\frac{\omega_{nom}(L_v + L_g)}{R_v + R_g}\right) > 65° \tag{2.22}$$

$$\left(\frac{\omega_{nom}(L_v + L_g)}{R_v + R_g}\right) > 2.14 \tag{2.23}$$

where R_g and L_g are the resistive and inductive grid components, and R_v and L_v are the resistive and inductive virtual impedance components. This case study assumes virtual inductive output impedance, neglecting the resistive impedance component.

The compensator parameters were tuned based on their characteristics. The proportional gain introduces a fast-transient response and it was set to a small value, since it can distort the signal rapidly, while the integral gain presents a slow and smooth response, however it is very sensitive, leading to a small gain as well.

2.3.2 Scheme and Subsystems

For the purpose of this case, the model designed is similar to the Case Study I scheme from Fig. 2.1, removing the AC power grid emulator and establishing the converter to work as voltage source.

As illustrated in Fig. 2.31, the scheme of the microgrid is composed by one network-forming converter modelled in series with one local load.

The control system is composed of the same three blocks as presented in the Case Study I, including additional variables to the Alpha_Beta block, illustrated in Fig. 2.32. The extra parameters correspond to the line current measurements, which are used to determine the reference voltage value that is applied in the voltage control

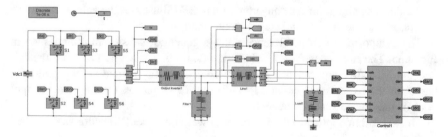

Fig. 2.31 Scheme of the microgrid composed by one network-forming converter and one local load

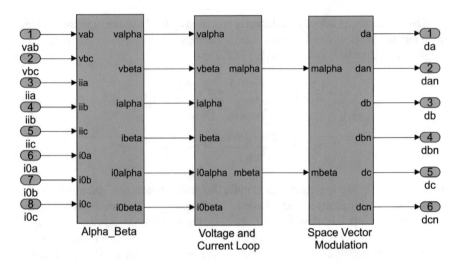

Fig. 2.32 Scheme of the control 1 subsystem

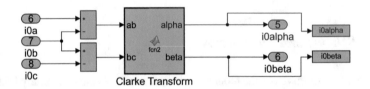

Fig. 2.33 Scheme of the αβ-frame subsystem

loop. These new variables are also converted to the αβ reference frame using Clarke transform as illustrated in Fig. 2.33.

Based on the converter measured values for the output voltage and the line currents in the network in αβ-frame, the voltage reference subsystem produces the reference voltages, defined as *valpharef* and *vbetaref*. These reference components are implemented in the voltage control loop to produce reference currents, which induces the modulating signal for the converter. Figure 2.34 illustrates the voltage reference generator, and the voltage and current control loops.

The composition of the reference voltage generator subsystem is presented in Fig. 2.35. The Reference Power Calculation (v, i0) block diagram includes the calculation for active and reactive power reference values, defined as *pref* and *qref*, based on the expressions defined in (2.8) and (2.9) and on the measured output voltage and line current components.

The algorithm applied to the reference power components function is described as follows:

```
function [pref,qref] = fcn2(valpha,vbeta,i0alpha,i0beta)

pref=(3/2)*(valpha*i0alpha+vbeta*i0beta);
```

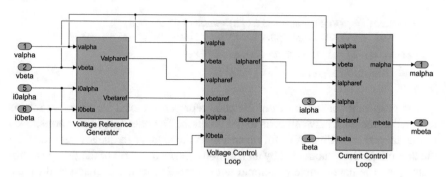

Fig. 2.34 Scheme of the voltage and current control loop subsystem

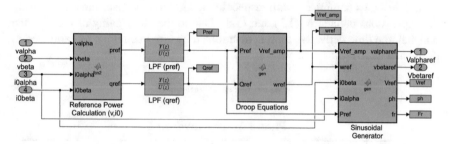

Fig. 2.35 Scheme of the voltage reference generator subsystem

```
qref=(3/2)*(-valpha*i0beta+vbeta*i0alpha);
```

In order to reduce distortions caused from the *pref* and *qref* produced signals, LPFs are implemented for each power component, and their transfer functions follow the expressions (2.11) and (2.12), stated in the Case Study I.

The Droop Equations block contains the MATLAB code applying the expressions (1.1) and (1.2), as following described:

```
function [Vref_amp, wref] = gen(Pref,Qref,mp, wnom, nq, Vnom)

Vref_amp = Vnom-nq*Qref;
Wref = wnom - mp*Pref;
```

The Sinusoidal Generator block produces the reference voltage as a function of time, which is characterized by the amplitude, phase angle *ph* and virtual impedance. The following MATLAB code describes the function.

```
function [valpharef,vbetaref,Vref,ph,fr] =
gen(Vref_amp,wref,i0beta,i0alpha,Lv,wnom,Pref,mpp,ts)
persistent ta;
if is empty(ta) ta=0; end;
t = ta+ts;
ph = (wref)*t-mpp*Pref;
if ph > (2*pi) t=((ph-2*pi)+mpp*Pref)/(wref); end;
```

```
valpharef  =  (Vref_amp)*sin(ph)+wnom*Lv*i0beta;
vbetaref  =-(Vref_amp)*cos(ph)-wnom*Lv*i0alpha;

Vref = sqrt(valpharef^2+vbetaref^2);

fr = (wref)/(2*pi);
ta = t;
```

The persistent constant *ta* from the code represents a local variable that retains its value in memory between calls to the function. The phase angle *ph* corresponds to the integral over time of (1.6), regarding to the angular frequency generated by the droop equations subtracted by an additional derivative term of the droop control technique. The phase angle *ph* should not exceed 2π, which is guaranteed by the MATLAB function by reducing 2π from the angle of the sinusoidal signal if the *ph* overpass its specified condition, as illustrated in Fig. 2.36.

The designed parameters with respect to the virtual impedance are derived from the expressions defined in (1.7) and (1.8). Due to the 90° lagging of the β to the α signal, the time derivative term is determined by cross-coupling, as expressed in (2.24) and (2.25) [6]:

$$Z_v i_{o\alpha} = R_v i_{o\alpha} + \omega_{nom} L_v i_{o\beta} \tag{2.24}$$

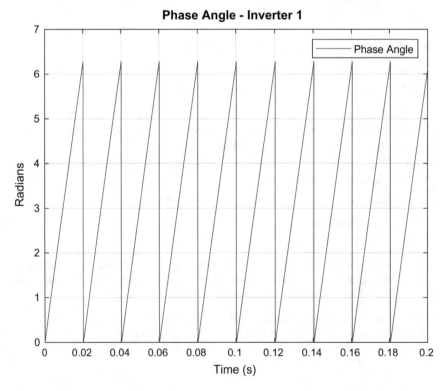

Fig. 2.36 Phase angle response

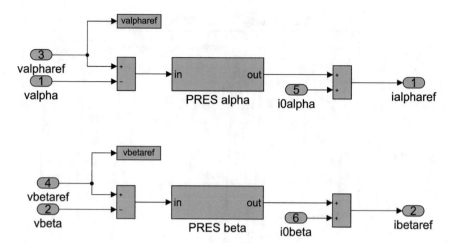

Fig. 2.37 Scheme of the voltage control loop subsystem

$$Z_v i_{o\beta} = R_v i_{o\beta} - \omega_{nom} L_v i_{o\alpha} \qquad (2.25)$$

In the voltage control subsystem, in order to eliminate the error comparing the output voltage and the generated reference voltage, a PRES is implemented. The output signal obtained from the PRES block is summed up with the network line current values in $\alpha\beta$ frame, called as feed-forward, to produce the reference current signals. Figure 2.37 illustrates the Voltage Control Loop subsystem.

The PRES transfer functions for the voltage control loop are presented in (2.26) and (2.27):

$$P_k(s) = k_{pv} \qquad (2.26)$$

$$RES(s) = \frac{k_{iv} 2shi_v \omega_{nom} s}{s^2 + 2shi_v \omega_{nom} s + \omega_{nom}^2} \qquad (2.27)$$

The reference current values produced by the Voltage Control Loop subsystem are supplied to the Current Control Loop block, which generates the modulating signals that are converted to pulses which drives the half-bridge converter, as explained in the Case Study I.

2.3.3 Simulation Results

The simulation results for the modulating signal, voltage and current tracking, active and reactive power, as well as the output voltage and instantaneous voltage are illustrated in Figs. 2.38, 2.39, 2.40, 2.41, 2.42 and 2.43.

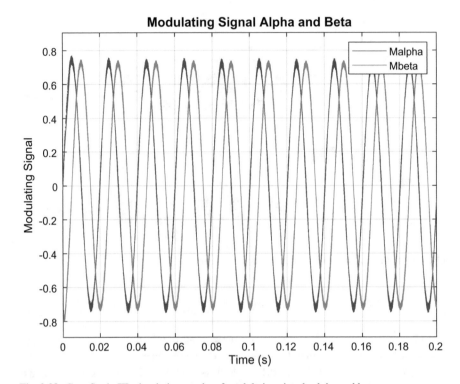

Fig. 2.38 Case Study III: simulation results of modulating signals alpha and beta

Figure 2.38 demonstrates the alpha leading the beta component by 90° ($^{\pi}/_{2}$), verifying accurate response for the modulating signal.

Figure 2.39 validates the current tracking of the generated alpha component over the reference value defined by the current control loop.

The voltage tracking from the voltage control loop shown in Fig. 2.40 demonstrates precise operation, based on the measured output voltage and line current components. Since the load is mainly resistive, the reactive power component present in Fig. 2.41 is nearly zero, while the active power corresponds to the estimated value in (2.1).

The result obtained by Fig. 2.42 reveals the stable operation of the system, which reaches the voltage nominal value. Figure 2.43 demonstrates that by applying only the primary control loop, the output frequency response presents a steady-state error that needs the secondary control loop for restoration purposes.

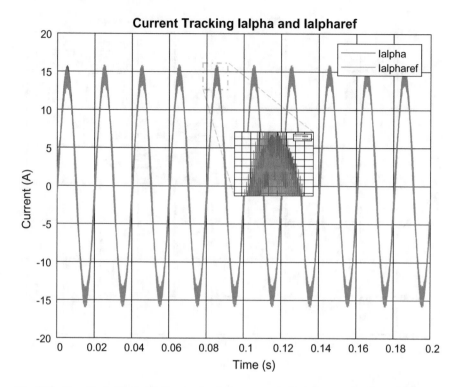

Fig. 2.39 Case Study III: simulation results of the current tracking from current control loop

2.4 Case Study IV: Islanded Mode—Microgrid Composed by Two Network-Forming Converters and One Common Load (Including Secondary Control)

2.4.1 Function and Control Parameters

Case Study IV contains two inverters operating as voltage sources, also called network-forming converters, and one common load. Additional to the droop-based primary control implemented in the Case Study III, it is applied a centralized secondary control to restore the voltage and frequency deviations caused by the primary control [11].

The electrical parameters for the additional inverter implemented in this Case Study are equivalent to 1.2 times the rated values specified for inverter 1, in order to emulate non-identical elements in the model. The values of the second inverter are described in Table 2.8. Additional to the control parameters applied in the previous case, Table 2.9 presents the adjusted parameters defined for the Case Study IV.

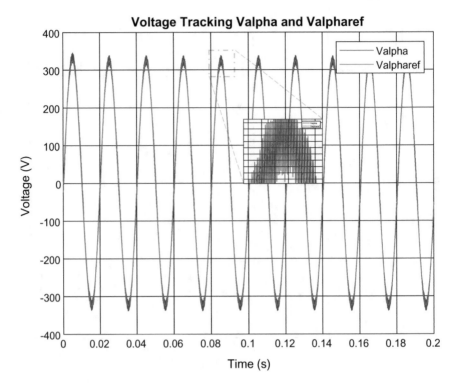

Fig. 2.40 Case Study III: simulation results of the voltage tracking from voltage control loop

2.4.2 Scheme and Subsystems

For the purpose of this Case Study, the model designed is similar to the Case Study III, including an additional converter, which will share the load with the former one after a specified period of time. The scheme of the microgrid composed by the two grid-forming inverters and the common load is presented in Fig. 2.44.

The inverters are arranged to connect to the microgrid progressively. While the inverter 1 starts operating at the initial instant, the inverter 2 starts operating after succeeding a defined period of time. The instant of time configured to trigger the second inverter is presented in Table 2.10. The inverter 2 is initially disabled, and after 2s plus t_s, it connects to the grid maintaining this operation mode up to a time sufficient to demonstrate the steady-state response.

Additional to the block diagrams presented in the previous case study, the inverter 1 will have a slightly different Sinusoidal Generator function, since it adds two new variables from an extra subsystem related to the secondary control. Figure 2.45 presents the voltage reference generator subsystem. Note that in this secondary control approach, the inverter 1 is the central (master) control node and the inverter 2 operates as a slave for this control loop.

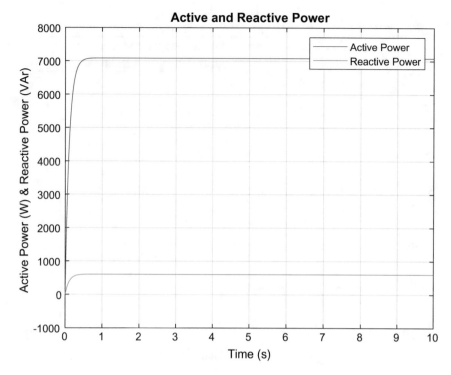

Fig. 2.41 Case Study III: simulation results of the active and reactive power

In the secondary control subsystem, as illustrated in Fig. 2.46, the voltage and frequency nominal values are compared to the measured ones. This control action is applied to decrease the voltage and frequency deviations, shifting them to their nominal values. The generated error is processed by a PI compensator at a sampling rate corresponding to 3 s plus t_s. The activation of the inverter by the secondary control is delayed by 1 s compared to primary control input, which is manually forced here to show the steady-state of the primary control loop before the secondary control is activated. This time delay is not necessary in a practical application and it is included here only for visualization purposes. The sampling sequence to enable the secondary control is detailed in Table 2.11.

The transfer function of the PI compensators, which are implemented in each of the loops to eliminate the error from the comparison between the nominal and measured values of the voltage and frequency, are described in (2.28) and (2.29), respectively:

$$PI_V(s) = \frac{k_{pV}s + k_{iV}}{s} \tag{2.28}$$

$$PI_f(s) = \frac{k_{pf}s + k_{if}}{s} \tag{2.29}$$

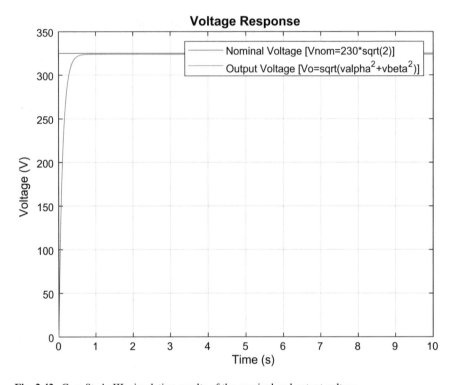

Fig. 2.42 Case Study III: simulation results of the nominal and output voltage

The compensators output signals are supplied to a sample/hold block, which samples continuously and freezes the value in a period of time. This block contains a memory output that retains the previous input value, as illustrated in Fig. 2.47.

The sampling rate adjusted in the sample/hold block is presented in Table 2.12. After the time constant t_s plus t_s, the sampling function is active at very small rate of 0.01 s, which improves the secondary control response.

The MATLAB code for the sinusoidal generation function introducing the deviation values calculated by the secondary control is described by the following function:

```
function [valpharef,vbetaref,Vref,ph,fr] =
gen(Vref_amp,wref,i0beta,i0alpha,Lv,wnom,dV,dw,Pref,mpp,ts)

persistent ta;
if is empty(ta) ta=0; end;
t = ta+ts;
ph = (wref+dw)*t-mpp*Pref;
if ph > (2*pi) t=((ph-2*pi)+mpp*Pref)/(wref); end;

valpharef = (Vref_amp+dV)*sin(ph)+Lv*wnom*i0beta;
vbetaref =-(Vref_amp+dV)*cos(ph)-Lv*wnom*i0alpha;

Vref = sqrt(valpharef^2+vbetaref^2);
```

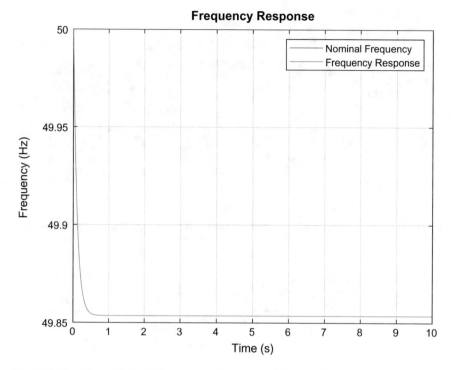

Fig. 2.43 Case Study III: simulation results of the output and nominal frequency

Table 2.8 Additional microgrid electrical parameters

Parameter name		Acronym	Value	Units
Inverter 2 I	Output inductance	L_{o2}	6	mH
Inverter 2 I	Output resistance	R_{o2}	0.6	Ω
Inverter 2 I	LC filter capacitance	C_{f2}	12	μF
Inverter 2 I	LC filter damping resistor	R_{f2}	48	Ω
Inverter 2 I	Line inductance	L_{L2}	1.2	mH
Inverter 2 I	Line resistance	R_{L2}	78	mΩ

Table 2.9 Additional microgrid control parameters

Parameter name	Acronym	Value	Units
Proportional gain PI voltage compensator	k_{pV}	0.1	A/V
Integral gain PI voltage compensator	k_{iV}	0.2	A^{-1}
Proportional gain PI frequency compensator	k_{pf}	0	A/V
Integral gain PI frequency compensator	k_{if}	10	A^{-1}

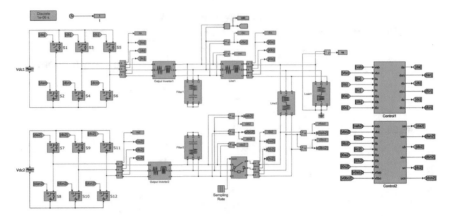

Fig. 2.44 Scheme of the microgrid composed by two network-forming converters and one common load

Table 2.10 Case Study IV: time to trigger the second inverter after $2 + t_s$ seconds	*Time values*			
	0	2	$2 + t_s$	1000 s
	Output values			
	0	0	1	1

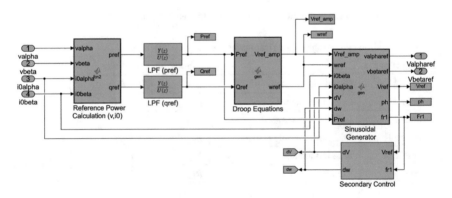

Fig. 2.45 Scheme of the control 1 voltage reference generator subsystem

```
fr = (wref+dw)/(2*pi);
ta = t;
```

The subsystem Control 2 includes additional variables to the Alpha_Beta block corresponding to the voltage measurements from the network, which are used in the PLL function, responsible for synchronizing the phase between inverter 1 and inverter 2. These new variables are also converted to the $\alpha\beta$ reference frame using Clarke transform as illustrated in Figs. 2.48 and 2.49.

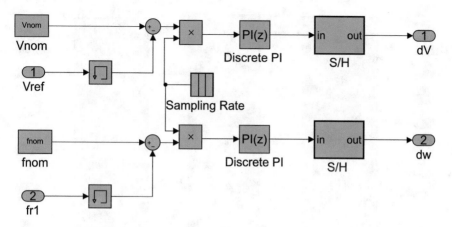

Fig. 2.46 Scheme of the secondary control subsystem

Table 2.11 Case Study IV: sampling rate to enable the secondary control at $3 + t_s$ seconds

Time values			
0	3	$3 + t_s$	1000 s
Output values			
0	0	1	1

Fig. 2.47 Scheme of the sample/hold subsystem

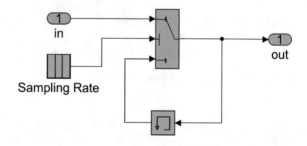

Table 2.12 Case Study IV: sampling rate to enable the sample/hold function at $t_s + t_s$ seconds

Time values			
0	t_s	$t_s + t_s$	0.01 s
Output values			
1	1	0	0

The synchronization role from the PLL is introduced in the Sinusoidal Generator function from inverter 2 by adding a phase increment variable *Aph* to the secondary control [6]. Figure 2.50 presents the voltage reference generator subsystem.

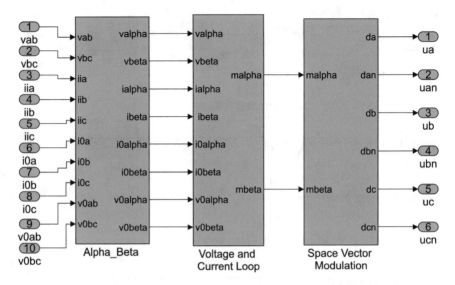

Fig. 2.48 Scheme of the control 2 subsystems

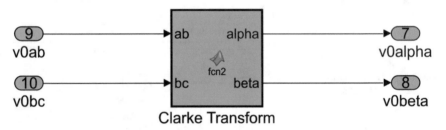

Fig. 2.49 Scheme of the αβ-frame subsystem

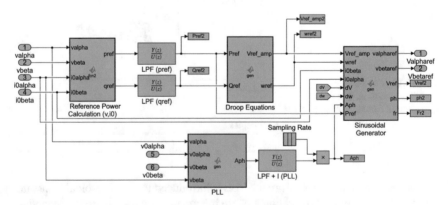

Fig. 2.50 Scheme of the control 2 voltage reference generator subsystem

The PLL MATLAB algorithm is based on the generation of a cross-product with reference to the sensed output voltage signal from inverter 1 and the voltage network signal measured from inverter 2, as following detailed:

```
Function Aph = gen(valpha,v0alpha,v0beta,vbeta)
Aph = valpha*v0beta - vbeta*v0alpha;
```

The *Aph* component produced by the PLL has its signal adjusted by a LPF, to attenuate harmonics and provide a slow dynamic for the new connected inverter, combined with an integrator compensator to minimize the phase angle difference concerning both voltage values [12]. The united transfer function of the LPF and I compensator is described by (2.30):

$$LPF\_I_{PLL}(s) = \frac{k_{iPLL}\omega_{cPLL}}{s^2 + w_{cPLL}s} \tag{2.30}$$

The sampling rate adjusted for the PLL subsystem enables the synchronization function with the reference voltage to start before the second inverter become active. Once the new inverter is synchronized with the network voltage and it is activated, the inverter is able to energize the microgrid and then collaborate to the power sharing function [3]. The sampling rate values are described in Table 2.13.

The Sinusoidal Generator MATLAB code introducing the synchronization value calculated by the PLL is described by the following function.

```
function [valpharef,vbetaref,Vref,ph,fr] =
gen(Vref_amp,wref,i0beta,i0alpha,Lv,wnom,dV,dw,Pref,mpp,ts,Aph)

persistent ta;
if is empty(ta) ta=0; end;
t = ta+ts;
ph = (wref+dw)*t-mpp*Pref;
if ph > (2*pi) t=((ph-2*pi)+mpp*Pref)/(wref); end;

valpharef = (Vref_amp+dV)*sin(ph+Aph)+Lv*wnom*i0beta;
vbetaref =-(Vref_amp+dV)*cos(ph+Aph)-Lv*wnom*i0alpha;

Vref = sqrt(valpharef^2+vbetaref^2);

fr = (wref+dw)/(2*pi);
ta = t;
```

The reference voltage values produced by the Sinusoidal Generator subsystem are sent to the Voltage and Current Control Loop subsystems, which generate modulating signals that are converted to pulses to drive the half-bridge IGBT switches, as explained in the previous case studies.

Table 2.13 Case Study IV: sampling rate to enable the PLL before the second inverter starts its operation

Time values			
0	2	$2 + t_s$	1000 s
Output values			
1	1	0	0

2.4.3   Simulation Results

The simulation results for the modulating signal, voltage and current tracking, active and reactive power, as well as the voltage and frequency are demonstrated in Figs. 2.51, 2.52, 2.53, 2.54, 2.55 and 2.56.

The modulating signals from Fig. 2.51 for both inverters provide an exact response by giving the alpha leading the beta component by 90° ($\pi/2$).

Figure 2.52 validates the current tracking of the produced alpha component over the reference value defined by the current control loop.

Figure 2.53 demonstrates precise voltage tracking from the voltage control loop, based on the measured output voltage and current components.

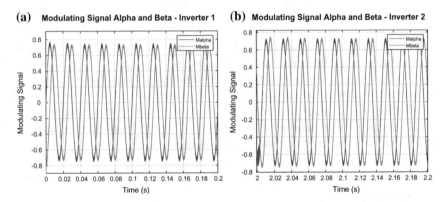

Fig. 2.51 Case Study IV: simulation results of the modulating signals alpha and beta, **a** inverter 1, **b** inverter 2

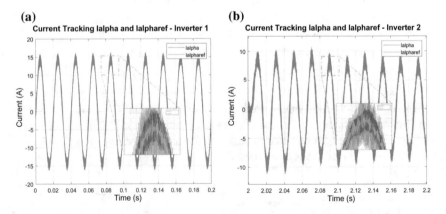

Fig. 2.52 Case Study IV: simulation results of the current tracking from current control loop, **a** inverter 1, **b** inverter 2

(a)   **(b)**

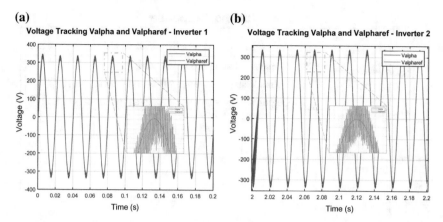

Fig. 2.53 Case Study IV: simulation results of the voltage tracking from voltage control loop, **a** inverter 1, **b** inverter 2

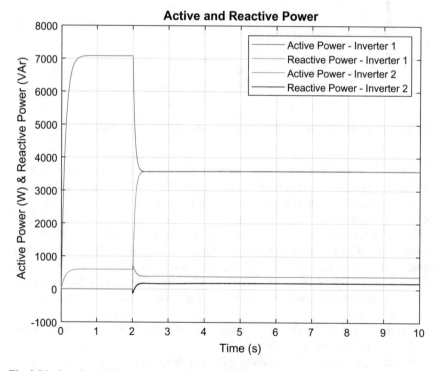

Fig. 2.54 Case Study IV: simulation results of the active and reactive power

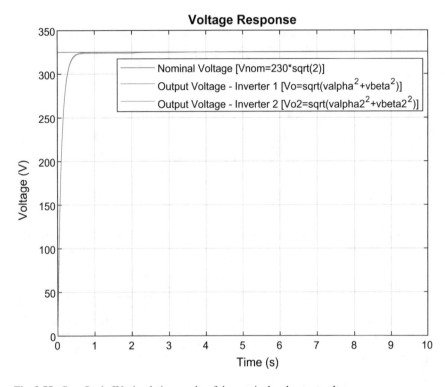

Fig. 2.55 Case Study IV: simulation results of the nominal and output voltage

Figure 2.54 presents the active and reactive power sharing between the two inverters into the microgrid, obtained by the droop control method. The inverter 1 starts operating by itself at t = 0 s providing the full required power, and at t = 2 s, the inverter 2 start sharing the power. As observed in the figure, the active power achieves a perfect power sharing, since the frequency is a global variable and all the nodes at the microgrid present the same frequency. In contrast, the reactive power sharing is not perfect achieved due to the fact that the voltage amplitude is a local variable that presents different values at each node. Note that these power sharing results were accomplished by considering perfect, lossless and without any communication network constraints.

The result obtained by Fig. 2.55 presents the stable operation of the system, which reaches the voltage nominal value.

Figure 2.56 validates the secondary control function of frequency restoration, providing an ideal transient response with soft start, mainly due to the PLL synchronization function applied with the LPF, and the time derivative term added in the droop control method.

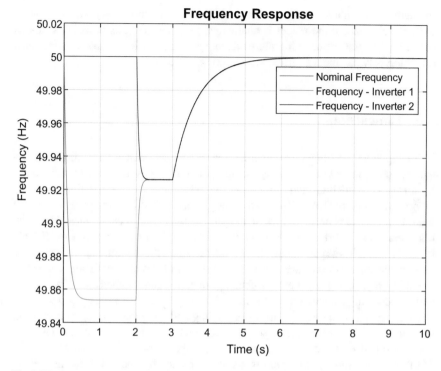

Fig. 2.56 Case Study IV: simulation results of the nominal and output frequency

2.5 Case Study V: Proposal for Transient Operating Mode—Microgrid Composed by One Network-Forming Converter, One Network-Feeding Converter and One Common Load

2.5.1 Function and Control Parameters

The Case Study V contains two inverters, one operating as a current source and the other operating as a voltage source, and one common load. The operation is based on both grid-connected and islanded mode. The principal objective of this case is to modulate variability on the grid operation profile, emulating scenarios as faults in the grid, leading to voltage sags and disconnection of the microgrid, allowing to observe the transition behaviour from the applied controls.

First the system should start operating connected to the grid, where the converters are operating in phase-angle synchronization to the grid performed by a PLL. At a predefined time, or due to some fault, the microgrid is disconnected from the grid at the PCC and the network-forming converter becomes the reference for the amplitude, frequency and phase signals. Islanding detection algorithms are needed to guarantee seamless transition between both operating modes [4]. This case is yet to be developed and the reader is challenged to investigate this practical scenario, searching into the available literature all the possible control schemes for a seamless transition between modes (grid connected or islanded).

2.6 Conclusion

This chapter has presented fundamental step-by-step guidelines for studies of different microgrid model scenarios regarding operation modes and control techniques, by means of modelling and simulating the systems using the Simulink software integrated with MATLAB. The methodology, results and performance analysis of microgrid control is presented when applying distinct control techniques to five distinct scenarios for grid connected, isolated and transient operating modes. For the first four cases, it was performed an analysis regarding the dynamic operations of the control, presenting the voltage, current and frequency transients, with both connected operating mode along with network-feeding converter, and islanded mode, along with network-forming converters. For the last case, a proposal study is described to assess transient operating mode with network-feeding and network-forming converter, in order to assess properties from disconnection and restoration processes among grid-connected and islanded mode. Analysis of simulation results have shown that the MATLAB-Simulink environment is a suitable choice for such projects, as well as demonstrating that the control techniques employed are effective in microgrids for sustaining power quality standards, and even stabilization of power flow against fluctuations in power generations typical of renewable energy sources.

References

1. Martí P, Velasco M, Martín EX, de Vicuña LG, Miret J, Castilla M (2016) Performance evaluation of secondary control policies with respect to digital communications properties in inverter-based islanded microgrids. IEEE Trans Smart Grid 9(3):2192–2202
2. Liserre M, Teodorescu R, Blaabjerg F (2006) Stability of photovoltaic and wind turbine grid-connected inverters for a large set of grid impedance values. IEEE Trans Power Electron 21(1):263–272
3. Miret J, García de Vicuña J, Guzmán R, Camacho A, Moradi Ghahderijani M (2017) A flexible experimental laboratory for distributed generation networks based on power inverters. Energies 10(10):1589
4. Guerrero JM, Chandorkar M, Lee T-L, Loh PC (2012) Advanced control architectures for intelligent microgrids—Part I: Decentralized and hierarchical control. IEEE Trans Industr Electron 60(4):1254–1262
5. Yazdani A, Iravani R (2010) Voltage-sourced converters in power systems, vol 34. Wiley Online Library, New Jersey
6. de Souza ACZ, Castilla M (2019) Microgrids design and implementation, 1st edn. Springer, Cham

7. Duesterhoeft W, Schulz MW, Clarke E (1951) Determination of instantaneous currents and voltages by means of alpha, beta, and zero components. Trans Am Inst Electr Eng 70(2):1248–1255

8. Sybille G, Dessaint L-A, DeKelper B, Tremblay O, Cossa J-R, Brunelle P, Champagne R, Giroux P, Gagnon R, Casoria S, Lehuy H, Fortin-Blanchette H, Tremblay O, Semaille C, Ouquelle H, Paquin J-N, Mercier P (2013) SimPowerSystems™ user's guide. Version 5.8 (Release 2013a) edn. Hydro-Québec Research Institute (IREQ) and The MathWorks, Inc., Natick, MA

9. Rey JM, Castilla M, Miret J, Camacho A, Guzman R (2019) Adaptive slope voltage control for distributed generation inverters with improved transient performance. IEEE Trans Energy Convers

10. Vasquez JC, Guerrero JM, Miret J, Castilla M, De Vicuna LG (2010) Hierarchical control of intelligent microgrids. IEEE Ind Electron Mag 4(4):23–29

11. Guerrero JM, Vasquez JC, Matas J, De Vicuña LG, Castilla M (2010) Hierarchical control of droop-controlled AC and DC microgrids—a general approach toward standardization. IEEE Trans Industr Electron 58(1):158–172

12. Rocabert J, Azevedo GM, Luna A, Guerrero JM, Candela JI, Rodríguez P (2011) Intelligent connection agent for three-phase grid-connected microgrids. IEEE Trans Power Electron 26(10):2993–3005

Printed in the United States
By Bookmasters